A Weekend in the City

Adding and Subtracting Times to the Nearest Minute

Colleen Adams

PowerMath™

The Rosen Publishing Group's
PowerKids Press™
New York

Published in 2004 by The Rosen Publishing Group, Inc.
29 East 21st Street, New York, NY 10010

Copyright © 2004 by The Rosen Publishing Group, Inc.

Book Design: Michael Tsanis

Photo Credits: Cover (buildings) © Ed Pritchard/The Image Bank; cover (clock) © Robert Daly/Stone; cover
(clock face) © EyeWire; p. 5 (present-day Chicago) © Bill Bachmann/Index Stock; p. 5 (Chicago 1850) ©
Bettman/Corbis; p. 6 © Mark Segal/Index Stock; p. 9 © Farrell Grehan/Corbis; pp. 10, 14 © Sandy
Felsenthal/Corbis; p. 13 © Jon Hicks/Corbis; p. 17 © Phillip Gould/Corbis; p. 18 © Richard Cummins/
Corbis; p. 20 © The Image Bank.

Library of Congress Cataloging-in-Publication Data

Adams, Colleen.
  A weekend in the city : adding and subtracting times to the nearest
minute / Colleen Adams.
        v. cm. — (PowerMath)
Includes index.
Contents: Chicago then and now — The Sears Tower — The Art Institute
of Chicago — Museum of Science and Industry — Grant Park — Shedd
Aquarium — Another Museum! — The Old Chicago Water Tower — Wrigley
Building — Going home.
  ISBN 0-8239-8974-7 (lib. bdg.)
  ISBN 0-8239-8897-X (pbk.)
  ISBN 0-8239-7425-1 (6-pack)
  1. Time measurements—Juvenile literature. 2. Chicago
(Ill.)—Description and travel—Juvenile literature. [1. Time
measurements. 2. Chicago (Ill.)—Description and travel.] I. Title. II.
Series.
  QB209.5.A33 2004
  529'.7—dc21
                              2003001262
Manufactured in the United States of America

Contents

Chicago Then and Now

My family and I are visiting Chicago this weekend. We plan to visit some **museums**, see some famous **landmarks**, and learn more about this great city. I have decided to keep a travel journal so I can keep track of how much time we spend at each place. My travel journal will also help us to remember everything we see and do.

I read that a man named Jean Baptiste Point DuSable first set up a **trading post** in the area now known as Chicago in 1779. Today, Chicago stretches for 29 miles along the shore of Lake Michigan and is the largest city in Illinois, with a population of almost 3 million people.

Chicago 1850

present-day Chicago

By 1890, there were over 1 million people living in Chicago. Today, Chicago is the third largest city in the United States.

Sears Tower

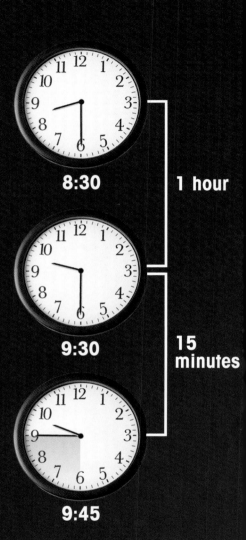

8:30   **1 hour**

9:30   **15 minutes**

9:45

1 hour and 15 minutes

The Sears Tower

June 5, 8:30 A.M.–9:45 A.M.

We finally arrived in Chicago! We left Union Station and took a taxi to the Sears Tower. We got there at 8:30. The Sears Tower is a **skyscraper** that was built in 1973. It is 1,450 feet high and has 110 floors. It is the second tallest building in the world and a famous Chicago landmark. We went up to the 103rd floor and looked out over downtown Chicago and Lake Michigan. When the sky is clear, you can see the states of Illinois, Indiana, Wisconsin, and Michigan from the top of the Sears Tower. We stayed until 9:45.

How long did we stay at the Sears Tower? There are 60 minutes in 1 hour and 30 minutes in 1 half hour. The clock face is divided into 5-minute sections. The dashes in between the numbers stand for minutes. If you count by fives around the clock from 8:30 to 9:45, you will see that we stayed for 1 hour and 15 minutes.

The Art Institute of Chicago

June 5, 10:10 A.M.*–12:30* P.M.

Next, we took a cab to the Art Institute of Chicago. We got there at 10:10. This famous art school and museum has a collection of over 300,000 pieces of art. We looked at different kinds of art from places around the world. Some pieces of art were very old and some were new. There was so much to see there, I lost track of time. When we left to go to lunch, it was already 12:30.

We left the art museum at 12:30. How much time did we spend there? Look at the clocks to find out.

The Art Institute of Chicago

| 1 hour | 1 hour | 20 minutes |

10:10   11:10   12:10   12:30

1 hour + 1 hour + 20 minutes = 2 hours and 20 minutes

Museum of Science and Industry

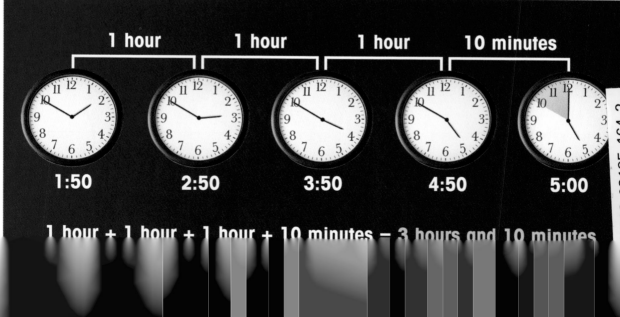

1 hour	1 hour	1 hour	10 minutes

1:50   2:50   3:50   4:50   5:00

1 hour + 1 hour + 1 hour + 10 minutes = 3 hours and 10 minutes

Museum of Science and Industry

June 5, 1:50 P.M.–5:00 P.M.

After lunch we took a cab to Hyde Park and visited the Museum of Science and Industry. We got there at 1:50. The museum is one of Chicago's oldest buildings. It was built in 1893 and was made into a museum in 1933.

At the Museum of Science and Industry, you can touch many of the objects that are on display. Our favorite things were a World War II submarine, a real coal mine that was built right in the museum, and a jet plane! There was so much to see that we stayed until 5:00.

How much time did we spend at the Museum of Science and Industry? Look at the clocks to find out.

Grant Park

June 5, 5:30 P.M.–6:50 P.M.

On our way back to our hotel, we visited Grant Park in downtown Chicago. There are many bridges, trees, and monuments in the park.

In the middle of Grant Park is the Clarence Buckingham Fountain, another popular Chicago landmark. The fountain is modeled after some famous fountains in Europe. In 1927, a woman named Kate Buckingham had this beautiful fountain built in honor of her brother Clarence. During the summer, the fountain is lit with lights of many different colors.

We ended our busy day at the Clarence Buckingham Fountain at 6:50. How long did we stay at Grant Park? Look at the clocks to find out.

Clarence Buckingham Fountain

1 hour   20 minutes

5:30   6:30   6:50

1 hour + 20 minutes = 1 hour and 20 minutes

Shedd Aquarium

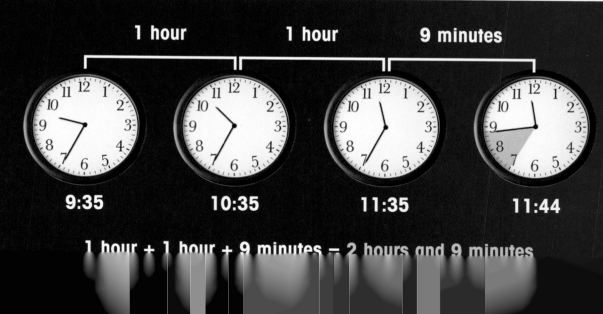

1 hour	1 hour	9 minutes

9:35   10:35   11:35   11:44

1 hour + 1 hour + 9 minutes = 2 hours and 9 minutes

Shedd Aquarium

June 6, 9:35 A.M.–11:44 A.M.

Today after breakfast, we went to another museum called the John G. Shedd Aquarium. It is the largest indoor aquarium in the world! This aquarium has over 6,000 sea animals from every part of the world. We saw whales, dolphins, seals, and seahorses. We saw a **coral reef** and a display that showed what the Amazon **rain forest** looks like. We took an hour for lunch. Then we decided to walk over to the Field Museum of Natural History.

Today we decided to visit 2 museums that are close to one another. How long did we spend at the Shedd Aquarium? Look at the clocks to find out.

Another Museum!

June 6, 12:55 P.M.–4:29 P.M.

The Field Museum of Natural History, which is located near the Shedd Aquarium, is one of the world's largest museums. The Field Museum has the fossil bones of Sue, the largest *Tyrannosaurus rex* fossil ever discovered. Sue stands about 13 feet high! Scientists think that Sue lived on Earth over 65 million years ago.

The Field Museum also has other displays about people and animals that lived long ago or in faraway places. We looked at displays about Egyptian **mummies** and Native Americans of North America.

How long did we spend visiting the Field Museum of Natural History? Look at the clocks to find out.

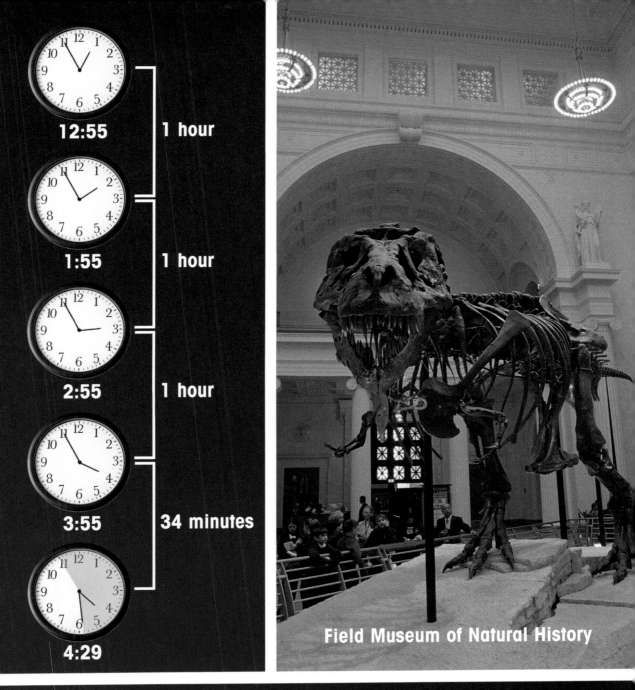

12:55 — 1 hour

1:55 — 1 hour

2:55 — 1 hour

3:55 — 34 minutes

4:29

Field Museum of Natural History

1 hour + 1 hour + 1 hour + 34 minutes = 3 hours and 34 minutes

Old Chicago Water Tower

If we get to the Old Chicago Water Tower at 9:35 and we spend one hour there, how much time will we have left to see another landmark before our train leaves at 12:35?

| 1 hour | 1 hour | 1 hour |

9:35   **10:35**   **11:35**   **12:35**

Total time before our train leaves = 3 hours
Time spent at Old Chicago Water Tower = 1 hour

3 hours – 1 hour = 2 hours

18

The Old Chicago Water Tower

June 7, 9:35 A.M.–10:35 A.M.

This morning we only had about 3 hours to sightsee before we had to catch our train home at 12:35.

First, we took a cab ride to the Old Chicago Water Tower. It is one of the city's oldest landmarks. The tower was built from limestone blocks in 1869. When a big fire burned most of the city's buildings in 1871, the water tower was one of the few buildings left standing. The tower is no longer used, but the water pumping station that is east of the towers still pumps water for the city. In 1969, the Old Chicago Water Tower was chosen to be the first American Water Landmark.

Wrigley Building

If we leave the Wrigley Building at 11:54, how many more minutes do we have until it is 12:00? There are 60 minutes in an hour, so we can subtract 54 minutes from 60 minutes to find out.

```
  5 10
  6̶0̶
-  54
_____
  6 minutes
```

6 minutes

11:54   12:00

The Wrigley Building

June 7, 11:00 A.M.–11:54 A.M.

The Wrigley Building on North Michigan Avenue was the last landmark we visited on our trip to Chicago. The Wrigley Building is named after William Wrigley Jr., who started his own company in 1919. First he sold soap, then he sold baking soda. Mr. Wrigley put a free stick of gum inside every package of baking soda. Soon, the gum became more popular than the baking soda! Mr. Wrigley's company grew and became very successful after he started selling only gum.

We spent 54 minutes at the Wrigley Building. Then it was time to go back to Union Station to catch the 12:35 train home.

Going Home

We want to go back to Chicago soon and see some of the sights we missed on this trip. I looked in my travel journal to figure out how much time we spent visiting landmarks and how much time we spent visiting museums. How much time did we spend altogether visiting landmarks and museums in Chicago? Looking at how we spent our time on this trip will help us plan for our next visit to the city.

	1
Time spent visiting museums:	**11 hours and 13 minutes**
Time spent visiting landmarks:	**+ 4 hours and 29 minutes**
	15 hours and 42 minutes

On this trip we spent a total of 15 hours and 42 minutes visiting museums and landmarks in Chicago.

Glossary

coral reef (CORE-uhl REEF) A chain of coral at or near the surface of the sea.

landmark (LAND-mark) A place that is important or interesting.

mummy (MUH-mee) A dead body that is prepared in a special way to make it last a long time. Some Egyptian mummies have lasted more than 3,000 years.

museum (myoo-ZEE-uhm) A building for displaying a collection of objects about subjects such as science, history, or art.

rain forest (RANE FOR-uhst) A very thick forest in a place where rain is very heavy throughout the year.

skyscraper (SKY-skray-puhr) A very tall building.

trading post (TRAY-ding POST) A store or station set up by a trader. At a trading post, people could get food, guns, clothing, and other items in return for animal furs and skins.

Index